THE ICE AGES

ROY A. GALLANT

THE ICE AGES

A FIRST BOOK

FRANKLIN WATTS
NEW YORK | LONDON | TORONTO
SYDNEY | 1985

Photographs courtesy of:
National Park Service: pp. 2, 13, 25, 26, 51, 52, 54, 55;
Science Photo/Graphics, Inc.: pp. 5, 29, 30, 44;
The American Museum of Natural History:
p. 16; U.S. Coast Guard: pp. 18, 19.

Diagrams courtesy of Vantage Art, Inc.

Library of Congress Cataloging in Publication Data

Gallant, Roy A.
Ice ages.

(A First book)
Includes index.
Summary: Describes the various ice age periods throughout
history and prehistory and explains how and why they occur.
1. Glacial epoch—Juvenile literature. 2. Glaciers—
Juvenile literature. [1. Glacial epoch. 2. Glaciers] I. Title.
QE697.G24 1985   551.3'1   84-19673
ISBN 0-531-04912-4

CONTENTS

—
—

FOR PAT

—
—

ACKNOWLEDGMENTS

The author wishes to thank
Macmillan Publishing Company for permission
to adapt, for inclusion in the present book,
certain passages from his book *Earth's
Changing Climate*, © 1979 by Roy A. Gallant.

My thanks also to Jeannine L. Dickey for
her valuable research assistance and for
operating the computer at various stages
in the preparation of the manuscript.

And finally, my thanks to George Kukla, of
the Lamont-Doherty Geological Observatory of
Columbia University, for checking the manuscript
for technical accuracy, and to Maury Solomon
for her customary thoroughness as my editor.

THE
"LITTLE ICE AGE"

A CREEPING WALL OF ICE

Try to imagine looking out your bedroom window and seeing a huge wall of ice—the front end of a **glacier**—near your house. Only the length of two city blocks away, the wall is 50 feet (15 m) high and 600 feet (180 m) wide—and it is moving, though very slowly, toward you.

This was the view that people living around the year 1600 had from their homes in the village of Chamonix, in the French Alps. The mountain glacier was advancing in response to a cold period called the **Little Ice Age** that gripped Europe from about the year 1400 to around 1850.

The Little Ice Age began in a mild manner. Increased precipitation—rain, hail, mist, sleet, and snow—in the mountains caused glaciers to grow. As a result, there was an average lowering of the temperature. Mid-latitude winters gradually grew a bit longer and became slightly colder. Summers were somewhat shorter and cooler than before. But the average lowering of the temperature was not anywhere near the 10°F (−12°C) that

Mountain glaciers form among the frigid peaks
of mountains where the snow accumulates faster
than it can melt. Two mountain glaciers are seen
here, one on either side of the central peak
of 18,000-foot-high Mount St. Elias, Alaska.

accompanies a major ice age. Instead, it was only a few degrees.

But that's all it takes to shorten the growing season by about two weeks in the midlatitudes, for example, in southern Maine and Wisconsin in the United States and France and Germany in Europe. At higher latitudes, such as where Canada, Iceland, and Norway are, the growing season is shortened even more. Advances of glaciers took place in 1600, 1640, 1740, 1810, 1820, and 1850 with resulting drops in average temperature. It is important to realize that even *small* average changes in climate over long periods of time are very significant. This is especially so in high latitudes, where the growing season tends to be short even in good times.

By the year 1600, several glaciers in the Alps had advanced and crushed houses, including some in the Chamonix Valley. Although a number of villagers abandoned their homes and moved elsewhere, others decided to stay and try to carry on, in hopes that the glacier would reverse its course. Many of these people starved to death, mainly because the shorter growing season prevented their crops from maturing. Those villages that survived the glacial advance of the Little Ice Age today have about half a mile (0.8 km) of woods and glacial rubble between them and the present forward wall of the glaciers.

Some winters of the Little Ice Age were especially fierce. The winter of 1709 was one of these. An Angers priest of France described the winter in these words: "The cold began on January 6th and lasted in all its rigor until the 24th. The crops that had been sown were all completely destroyed. . . . Most of the hens had died of cold, as had the beasts in the stables. When any poultry did survive the cold, their combs were seen to freeze and fall off. Many birds, ducks, partridges, woodcock, and blackbirds died and were found on the roads and on the thick ice and frequent snow. Oaks, ashes, and other valley trees split with cold. Two-thirds of the walnut trees died."

The year 1739 brought another fierce winter to Europe. Belgium was especially hard hit. According to diaries left by many people, there was no spring season that year, and bad weather lasted well into May. The summer was short, cold, and wet. A late wheat harvest was spoiled by rain; wine grapes and other fruit harvests were destroyed by early frosts. Food was in such short supply that many of the poor rioted.

Although most winters of the Little Ice Age were not as severe as those of 1709 and 1739, they were more severe than the winters we have now. People living in Iceland and the Scandinavian countries were hit even harder than people living farther south. The Icelanders had been mainly a grain-growing people since about the year 900. But with the onset of the Little Ice Age, they were able to grow grain only in the southern regions of their country. And finally, of the different kinds of grains they had grown over the centuries, only barley could survive the cold. By the year 1500 the Icelanders had completely given up attempts to grow grain at all, even barley.

THE GULF STREAM
CHANGES COURSE

Another event that worsened the fate of the Icelanders during the Little Ice Age was a change in the course of the **Gulf Stream** beginning sometime after 1600. The Gulf Stream is a "river" of warm water that flows out of the Gulf of Mexico and toward Europe in a northeasterly direction, going up the North American coast. It crosses the Atlantic Ocean south of Newfoundland and flows toward Europe. As it does, it breaks into two major branches, one flowing northward toward the British Isles and the other flowing southward. The northbound branch then branches again, one arm flowing up toward Iceland.

The warm waters of the Gulf Stream had provided Iceland, Britain, and parts of Scandinavia with a moderate climate. But then the Gulf Stream gradually began to shift its transatlantic

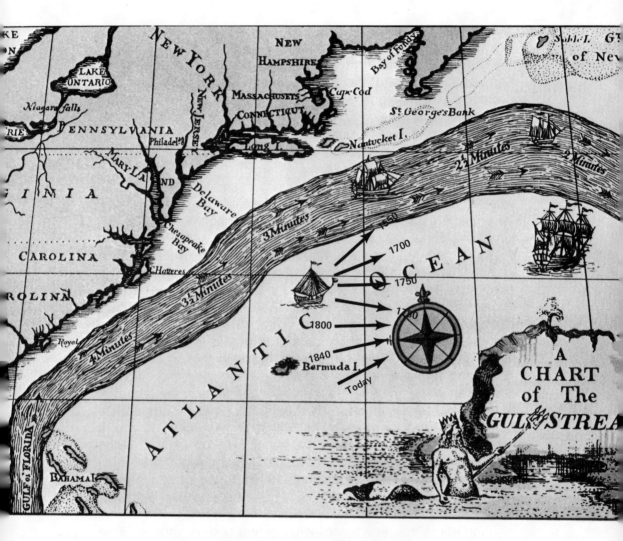

This historical chart of the Gulf Stream,
prepared by Benjamin Franklin, shows how this
warm river of water flowing out of the Gulf of
Mexico and across to Europe has changed course
from time to time since the year 1550.

course from northeast to east, and by about 1780 it was flowing in a slightly southeasterly direction.

The effect of this shift was to push the warm waters of the Sargasso Sea area of the Atlantic Ocean south, permitting cold Arctic water to move farther south. The result was a southward invasion of drift ice around Iceland, cooler summers, and colder winter winds. It wasn't until well into the 1900s that the Gulf Stream resumed its nearly northeast course of four centuries earlier.

The United States seems to have been less hard hit than western and northern Europe during the Little Ice Age. Because the Indians of North America did not keep written records, we know less about the climatic conditions of North America early in the Little Ice Age than we do about conditions in Europe. However, scientists who study past climates around the world say that after about 1400 weather conditions in the United States became cooler and/or wetter. (See Chapter 7 to find out how scientists can date events in the Earth's past.) Records kept by the army in the mid-1800s show that temperatures during that period were about 3.5°F (1.9°C) lower than over a similar period during the 1930s and 1940s. In regions that tend to have short growing seasons even when times are favorable, a decrease in the length of the growing season by about a month can have serious consequences in food production. As Maine farmers can tell you, a drop in temperature of only a degree or two can mean a killing frost early in September and the loss of several vegetable crops. Still other records for this same mid-1800s period show a tendency toward more northerly winds, which helped keep temperatures low.

With some appreciation of what only a two- or three-degree drop in the average temperature can do, you can begin to imagine the results of a drop in the average temperature of more than twice that amount—a major ice age that covers the land with thousands of feet of ice.

2

AGES OF ICE

ICE, ICE, AND MORE ICE

Climatologists, scientists who study climate, tell us that during more than 90 percent of the past 570 million years of Earth's 4.6-billion-year history, the north and south polar regions have probably been entirely free of ice. Also, during most of this time, the average world temperature was probably about 72°F (22°C). Palm trees once grew in most areas of what is now the United States. New York State had a climate like the one Florida has today. But ages of ice were to interrrupt this semitropical setting. For example, there was a glacial period about 500 million years ago, and another that lasted for some 50 million years about 300 million years ago. After that the Earth remained free of ice ages for about 250 million years, until the time in geologic history called the **Tertiary period.**

The Tertiary period began about 65 million years ago and brought with it a new series of ice ages—times when sheets of ice thousands of feet thick covered vast stretches of the land. Scientists cannot say for certain exactly what caused this most

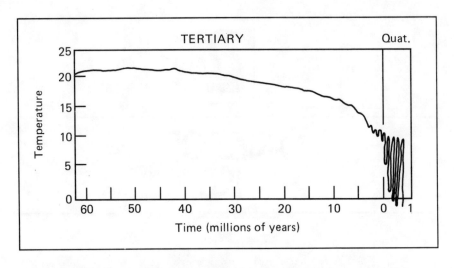

During the Tertiary, midlatitude temperatures gradually declined, although temperatures in the tropics seem to have remained unchanged. The suspected cause of the gradual decline at midlatitudes was the shifting of Antarctica to its present position.

recent series of outbreaks of ice, but they have developed some plausible theories.

During the Tertiary period, there seems to have been a gradual drop in temperature along the midlatitude in the Northern Hemisphere. This is about halfway between the equator and the North Pole. One possibility for this ancient period of cooling was the gradual movement of the Antarctic continent to its present position at the South Pole. Some 200 million years ago there seems to have been a single huge continent that scientists call **Pangaea**. Later, Pangaea broke apart into two landmasses, **Gondwana** in the south and **Laurasia** in the north. These landmasses in turn later broke apart into the continents we know today and drifted to their present positions.

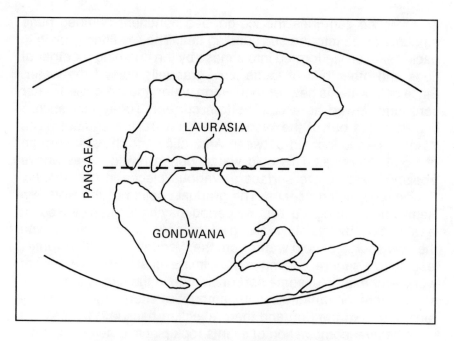

*It now appears that soon after the Earth
first established a solid crust there may have
been just one mammoth continent, called Pangaea.
It later broke apart into two large land masses,
Gondwana in the south and Laurasia in the north.
These land masses in turn later broke apart into
the continents we know today.*

By the end of the Tertiary the Antarctic continent had drifted to the region of the South Pole. Winter snows would have gradually built up over the following centuries and become packed into ice. Eventually, ice would have covered the entire continent—as it does today—to a depth of about 10,000 feet (3,000 m). Antarctica has been buried beneath its thick caps of ice for the past several million years.

Over the centuries the ice-gripped continent dumped huge amounts of ice into the surrounding sea. The resulting growth in pack ice—sea ice formed into a mass by the crushing together of floes and other bits of loose ice—gradually cooled the water. Since cold water is heavier than warm water, the ice-chilled water sank and flowed as a cold bottom current. Today just such a current flows out of the Weddell Sea toward the equator. (The Weddell Sea is located between Antarctica and the southern tip of South America.) This cold water, only a few degrees above freezing, rises to the surface at various places in the Atlantic, Pacific, and Indian oceans. The gradual cooling of the Northern Hemisphere during the Tertiary period may have been caused, at least in part, by this slow mixing of the oceans' warm water with the new source of cold water from the Weddell Sea. Other forces leading to a general cooling of the climate must also have been at work—for example, some sort of change in the circulation of the atmosphere, or variations in the amount of energy the Earth was receiving from the Sun and then reflecting back into space.

Whatever combination of events took place, between about 2 and 3 million years ago, mountain glaciers began to form in the Sierra Nevada of California and in Iceland. Climatologists are not certain about the timing of the ice buildup on Greenland. In any case, Greenland also become covered with about 10,000 feet (3,000 m) of ice. When it did, it became the source of a cold-water current that helped cool the waters of the North Atlantic ocean.

THE BIG MELT
13,000 YEARS AGO

Climatologists agree that several glacial periods have come and gone over the past few million years. Over the past 700,000 years there have been seven known glacial periods, each separated from the next by a period of warming called an **interglacial period.** Those of us presently living in the Northern Hemisphere may be enjoying the peak of one such warm, interglacial period now.

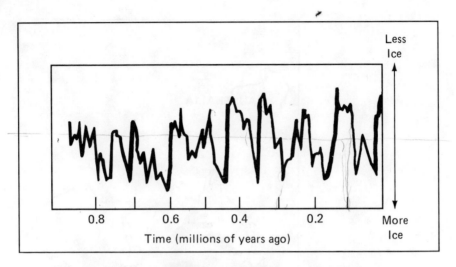

Over the past 700,000 years there have been seven known glacial periods, each followed by a period of warming (heavy lines). The most recent period of warming (heavy line at far right) has brought us to a stage where it is now warmer than it has been for more than 90 percent of the past million years.

According to climatologist Reid A. Bryson, "To find a time as warm as the past few [thousand] years, we have to go back through a long glacial period to 125,000 years ago." Each cycle of glacial activity—from the peak of one glacial period, through an interglacial period, then to the peak of the next glacial period—lasts about 100,000 years. From the end of one glacial period to the beginning of the next lasts about 10,000 years.

The peak of the last glacial period came about 18,000 years ago. It ended about 10,000 years ago. At the peak, ice covered about 30 percent of the Earth's total land surface, and it formed in both the Northern and Southern hemispheres. In the Northern Hemisphere, and in northern Labrador and in central Norway, for

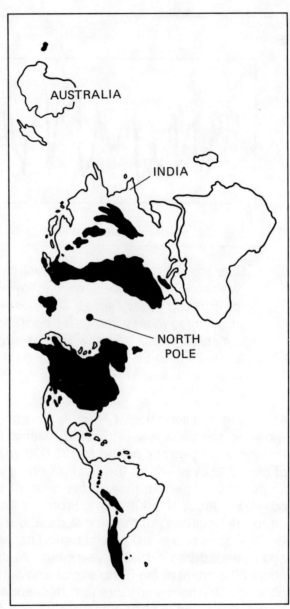

AUSTRALIA

INDIA

NORTH POLE

At the peak of the last glacial period, some 18,000 years ago, ice (dark regions in the diagram) covered about 30 percent of the Earth's land surface.

An enormous mountain glacier, the Ruth Glacier,
flows along a 25-mile-long valley from the area
near the peak of Mount McKinley

example, the ice probably first formed into glaciers in the mountain regions. These so-called **mountain glaciers** then grew into sprawling ice sheets that gradually crept south, covering the land with several thousands of feet of ice. Although the average thickness of ice was about 4,000 feet (1,200 m), in some places the maximum thickness probably reached 13,000 feet (3,900 m), or more than 2 miles (3.2 km).

During a glacial period it just snows and snows. But it also rains, although more water than today is deposited on land as snow. Ocean water evaporates and rises into the atmosphere, where it turns to snow and rain. In this way increasing amounts of ocean water are evaporated, with the result that the sea level drops significantly during an ice age. During the last glacial period the ocean level may have dropped about 300 feet (90 m). Although the polar oceans are especially cold and largely ice-covered during a glacial period, the ocean water in low latitudes and near the equator remains warm. During the last ice age of 18,000 years ago the Caribbean Sea, for example, was a balmy 73°F (23°C) compared with 80°F (27°C) today.

3

AS GLACIERS ADVANCE

HOW A GLACIER GROWS

Scientists who study glaciers, called **glaciologists**, can observe the ice and put together the chain of events that take place during an ice age, the process of accumulation and spreading of ice as a glacier forms and then fans out. This is just what is happening today in Greenland, for example. During storm after storm the snow accumulates in thin layers, first in the mountainous regions. The snow layers gradually become compacted. These compacted layers, called **firn**, later turn to ice, the bottom layers of which flow and spread ever downward and out toward the growing ice sheet's edges.

Perhaps you are wondering how ice, which is a solid, can *flow*. Especially large accumulations of ice, such as those making up a glacier, weaken the crystal form of the ice that makes a smaller piece of ice rigid. Thus, glacial ice tends to be softer and, without melting, can flow. Glaciologists drill holes into a glacier, then stick pipes into the holes. By measuring the changes in the positions of the pipes weeks or months later, they can tell how fast this or that part of the glacier is moving.

Mountain glaciers carry rock, soil, and other materials along, flowing toward the sea, meandering as they follow the steepest incline toward sea level.

Cold glaciers, the temperatures of which are below the melting point of ice, tend to flow over the bedrock more slowly than warm glaciers, the temperatures of which are at the melting point. A typical glacier flowing down a valley, called a **valley glacier**, may move about 650 feet (200 m) a year, though some glaciers have been known to move about 3 miles (4.8 km) in one year. A few called "galloping glaciers," move in huge spurts and travel as much as 4 miles (6.4 km) in only a few months.

The ice of a flowing glacier spreads outward over hundreds of miles. At first it follows routes set by the shape of the land. As the glacial ice flows through narrow regions, it is squeezed and speeded along. On reaching the shore, a glacier like this is called an **outlet glacier**. In Norway and Alaska, among other places, outlet glaciers have gouged deep, narrow valleys out of the bedrock leading into the sea. These gouged-out valleys are called **fjords**.

HOW ICEBERGS ARE BORN

Along other regions of the coast, the advancing ice sheet reaches the edge of the land and continues on, pushing its way into the water. As it does, the end of the ice sheet resting in the water is buoyed up. The resulting pressure causes a large section of ice to snap off along those zones of weakness, which are called **crevasses**. The breaking-off process is called **calving**. The resulting blocks of free-floating ice, often enormous, are **icebergs**. Most such bergs are the by-products of the fast-moving outlet glaciers. Some scientists estimate that the Greenland ice sheet today may release as much as 50 cubic miles (80.50 cu km) of ice into the sea each year. According to geologist James L. Dyson, the Rink Glacier alone, on Greenland's west coast, "has dumped an estimated 500 million tons of ice into the sea in only a few minutes— and it repeats this feat about once every two weeks."

From 10,000 to 15,000 large icebergs calve into the sea from Greenland's shores every year. In 1943, U.S. Coast Guard ice-

Icebergs are said to "calve" off the ends of glaciers leading to the sea. The large berg that just broke off the end of Jacobshavn Glacier, Greenland, plunged into Baffin Bay. It is about five miles wide. Many of the icebergs shown here are equal in size to several city blocks.

*Icebergs like this giant, photographed
by the U.S. Coast Guard in Baffin Bay,
are reminders that from time to time our
planet is gripped by centuries of intense
cold that we call an ice age.*

berg census takers counted a total of 40,232 bergs. Some of Greenland's bergs are nearly 1 mile (1.6 km) long or longer and tower 300 feet (90 m) above the ocean surface. The one shown here is 1,000 feet (300 m) long and 400 feet (120 m) above the water line, but its ice extends to a depth of 950 feet (290 m) below the ocean surface! So in order to recreate the last ice age at its peak, we need only visit Greenland today.

If the Greenland **icecap** were to melt suddenly, so much water would be added to the oceans that the sea level would rise about 25 feet (7.6 m). But "sudden" melting is not the rule among glaciers. Some experts think that it would take a period of warming typical of the temperature rise over the past 100 years for a period of about 10,000 years to melt all the polar ice. Others disagree, saying that the time needed would be only a few thousand years.

In North America during the last major glacial period just about all of Canada was covered with ice, and the eastern United States was covered as far south as 38° to 39° north latitude, near what is now St. Louis, Missouri. Perhaps because of relatively little snowfall, the lowlands of northern Alaska and parts of the Canadian Archipelago did not accumulate glaciers. The mountain glaciers that formed in southern Alaska and in the Pamir-Hindu Kush-Karakoram part of central Asia did not grow into large ice sheets. In nearly all of the high mountain areas the level of the permanent snow lines crept lower down the mountain slopes by about 4,500 feet (1,400 m). Mountains along the tropics also had their permanent snow lines lowered by approximately 3,000 feet (900 m).

In the Southern Hemisphere the ice was confined mostly to the mountain regions and to countries in high latitudes such as New Zealand. On land—especially in Africa and in South America—grasslands, steppes, and deserts spread at the expense of forests because of a general drying. Coupled with the sprawling ice in both hemispheres, this change in vegetation caused a greater amount of the Sun's energy to be reflected back into space than is being reflected back today.

We might at first think that these massive blankets of ice would cause a sharp drop in the average world temperature, but this is not the case. The average drop in mean (average) temperature from the previous interglacial period to the last major ice age was only about 10°F (5.5°C), but some regions of the mid-latitudes and somewhat higher latitudes had even greater drops in their mean temperatures. For example, the temperature change in Greenland was about 20°F (11°C). This sharp drop was due partly to higher elevation and partly due to air flowing down over the surface of Greenland's great ice dome and being rapidly cooled. Coastal Ireland, France, and northern Spain had similar drops. In contrast, the temperature drop along the Pacific Coast of North America, at the same latitudes, was closer to the average of 9°F (5°C). This marked difference between the Pacific and Atlantic regions seems to have been caused by the initial lowering of sea level at the 130-foot (40-m) stage. When the Pacific Ocean was lowered by that amount, the Bering Strait between Alaska and Siberia was closed. This meant that relatively warm water from the Pacific could no longer flow into the Arctic Ocean. However, the flow of cold Arctic water down into the Atlantic continued.

BIG ICE CHANGE—
SMALL TEMPERATURE CHANGE

Increased precipitation in the mountains causes an increase in mountain glaciers, which advance and eventually may bring about a change of 3° to 4°F (1.6° to 2.2°C) in average world temperature and initiate an ice age. Reduced precipitation can reverse the trend, turning a glacial period into an interglacial period accompanied by a rise in the average world temperature by about the same amount.

We might think that such widespread ice coverage as occurred during the last glacial period would have caused great harm to most plants and animals and lead to the extinction of many of them. But apparently this was not the case. One important effect of the advancing and retreating ice was a change in the *distribution* of plant and animal populations. For example, glacial deposits found in central and western Europe contain the remains of woolly rhinoceroses, mammoths, lemmings, reindeer, Arctic foxes, and moose—all species adapted to cold-weather environments. However, interglacial deposits in the same regions contain the remains of forest elephants, deer, rhinoceroses, hippopotamuses, and wild pigs—all adapted to a warmer climate, a climate characteristic of Africa today. In New England during the last period of glacial retreat reindeer and woolly mammoths roamed the land and moose ranged over the New Jersey countryside.

At the peak of the last glacial period, as commonly occurs, there was a dry period. Possibly such periods are caused by the cooler surface water of the oceans and an accompanying decrease in evaporation, which would result in smaller amounts of water vapor in the air. The evaporation rate may drop by 30 or 35 percent at such times. During these periods of dryness the winds pick up large amounts of fine dust, called **loess**, which slowly rain down on the Earth over long periods of time.

Now that we have some understanding of what an ice age is and what happens to the land and sea during an ice age, let's put together some of the pieces of the geological puzzle left by a period of widespread glaciers.

4

TRACKING THE GREAT ICE

THE WAVE OF DOOM

At the narrowest part of the Alaskan panhandle is Lituya Bay, which is sometimes used by fishermen as a safe place to anchor their boats. But the three small boats that put into that shelter during the night of July 9, 1958 would have been much better off in a storm-tossed sea.

Lituya Bay is an ice-carved fjord about 2 miles (3.2 m) wide and reaching some 7 miles (11.2 km) inland, where it receives a flow of glacial ice and rock debris from a much larger and nearly glacial-filled fjord.

Around 10 o'clock on that fateful July night, an earthquake sent 90 million tons of rock crashing down onto the Lituya Glacier at the head of the ice-choked fjord. The falling rock sheared off more than 1,000 feet (300 m) of the glacier. When the rock debris and ice plunged into the water it raised a mountainous wave 1,800 feet (540 m) high! The wave raced down the fjord toward the three tiny boats at anchor. Along the way it stripped away forests, soil, and everything else in its path right down to the bedrock. The ear-splitting crash of the rock and ice slide had alerted the fishermen that something very unusual was happening. As they started their engines in hopes of fleeing, they were horrified to see bearing down on them a monstrous wave. Although mostly spent, the wave was still about 100 feet (30 m) high and moving about 100 miles (160 km) an hour. One boat was instantly demolished and vanished without a trace. Another was flung a quarter of a mile through the air and smashed to splinters among the log debris carried down by the wave. The third boat, with fisherman Howard Ulrich and his seven-year-old son, miraculously survived. The boat was lifted up by the wave front onto the wave's crest high above the treetops along the shore. It was then lowered into the log-filled churning waters of the bay. Ulrich and his son were too stunned to even wonder what had happened. They were simply happy to be alive.

READING THE SIGNS OF A GLACIER

Glacial ice left over from the last great ice age still covers much of the land. In its relentless march toward the sea, the ice continues to shape mountain peaks now completely or partly buried beneath the ice. It gouges out enormous valleys, and generally carries out earth-moving feats that defy imagination. Today's glacial ice is found on every continent except Australia and covers about 10 percent of the Earth's total land area. The continental glacier that caps Antarctica has enough ice to cover the United States to a depth of 9,000 feet (2,700 m). All this ice is a frigid reminder of the great North American ice sheets that last withdrew 10,000 years ago.

Knowing how to read the telltale signs left by an advancing or retreating glacier helps convey an impression of the awesome force of a multimillion-ton mountain of ice on the move. A valley glacier, like a mammoth ice sheet, grows when its accumulation of snow and ice is more rapid than the loss through melting. Because the ice at the base of a glacier is under great pressure

The Yetna Glacier on Mount McKinley shows the formation of lateral moraines. These are long, dark bands formed as the glacier dislodges from the valley walls boulders and other rubble that tumble onto the ice. Here two valley glaciers join, their lateral moraines forming a single mid-moraine called a medial moraine.

A grand view of Yosemite Valley,
the bedrock floor of which lies buried
beneath up to 1,800 feet of ice.

from the weight of the ice above, it flows downhill as it is forced outward from its source. As we saw in the previous chapter, the speed of glaciers varies. One of the fastest glaciers on record is India's Kutiah Valley Glacier, which flowed 370 feet (113 m) a day for a three-month period in 1953.

Large or small, an advancing glacier bulldozes the surface, cutting, grinding, and carrying away the loosened rock rubble, called **glacial drift**. As a mass of ice flows down a narrow river valley it acts like a rasp file and scrapes deeply into the valley floor. It also cuts away and broadens the valley walls, causing boulder-size rocks and other rubble to tumble onto the ice margins, where they accumulate as long, dirty bands of rubble called **lateral moraines**. Gradually, the rocks tend to sink into the ice. Where two valley glaciers meet, they flow together as one, and their lateral moraines form a single flow, or **medial moraine**. Farther down the mountain still more valley glaciers may join the main flow until there are several long, curving medial moraines flowing down the valley. These valley glaciers can be huge. In Alaska and the Himalayas there are some that are 70 miles (113 km) long and 3,000 feet (900 m) thick. The Yosemite Valley is an old and typical U-shaped glacial trough, as opposed to the V-shaped canyons and valleys scoured out by rivers. The Yosemite Valley's bedrock floor lies buried beneath up to 1,800 feet (540 m) of clay and sand, deposited there when, for a while, the valley was a lake formed by the glacier's melting ice. The valley of the Roaring Fork River in Colorado is also an old glacial trough.

The rocks, sand, and other debris carried along by ice at the bottom of the glacier act like a very coarse sandpaper. The finest of these so-called **glacial tools** are sand and silt. They polish the bedrock smooth. Larger pieces leave small parallel scratches called **striations**. Still larger pieces may leave grooves the width of a thin rope. All this action of a glacier grinding its way down a valley produces a fine silt called **rock flour**. Sometimes enormous boulders, called **erratics**, are carried many miles by a glacier on the move and eventually are deposited in a region that does not have rock types like the erratic. These then are features of the bedrock.

The forward end of a glacier can no longer advance when warming temperatures cause the rate at which it melts to equal the rate at which the flow ice is replaced. When the melting rate is greater than replacement, the glacier begins to retreat. Meanwhile, replacement ice dumps new loads of drift material onto the original load. As the ice retreats it releases its drift debris in a series of low hills. This large area of jumbled silt, clay, sand, gravel, and boulders—all carried from sources many miles away—is called a **terminal moraine**. An especially impressive sprawl of glacial debris can be seen by walking across the lower part of Nisqually Glacier on Mount Rainier in Washington. Tons of glacial drift completely cover the ice in places, and tons more choke the valley below the glacier's forward edge. Another terminal moraine forms the backbone of Cape Cod in New England.

Meltwater from the end of the glacier forms streams that carry off and sort the finer materials of the glacial drift. The rock flour turns these meltwater streams a milky color. Where a glacier has retreated, its terminal moraine may collect a covering of soil and vegetation over the centuries. The old glacier may then advance again and deposit a new terminal moraine on top of the old one. In road cuts along highways you can sometimes identify successive terminal moraines stacked one above another and separated by thin layers of soil formed during the interglacial periods. Like mountain glaciers, the vast North American ice sheets that gouged out the Great Lakes and New York's Finger Lakes piled up millions of square miles of moraines. During the several hundred thousand years that it took to grow, that great ice sheet scraped an average of 30 feet (9 m) of rock off the land in one of the Earth's greatest face-lifting operations.

There are other telltale signs left by a glacier that once passed our way. Agates, a variety of the mineral quartz, found in glacial debris near Topeka, Kansas, have been identified as com-

As a glacier flows over the bedrock, sand in contact with the bedrock polishes the rock smooth while coarse gravel grinds out small parallel scratches called striations.

Boulders often are carried many miles from their source by a glacier on the move. As the ice melts, the boulders, called erratics, are deposited and can be identified when they are different rock types from the local rock.

ing from rocks along Lake Superior. The Topekan glacial sediments must have been carried more than 600 miles (965 km) by the great ice sheet. Diamond fragments found in glacial sediments of the Great Lakes states are scattered in such a way that they point to a potentially rich diamond source somewhere near James Bay in Canada. When the source of such glacier-carried materials is known, the materials become indicators that enable glaciologists to trace the flow path of a past glacier.

KETTLE LAKES, ESKERS, AND DRUMLINS

Sometimes the retreating front of a glacier leaves behind an area of debris-covered ice many square miles in size. Detached from its parent glacier and well insulated from solar heating, this ice stagnates for centuries before melting completely. Weak spots in the stagnating ice sometimes collapse under heavy debris loads and form distinctive features called **kettle lakes**. Hollows may also form in the blanketed ice. Meltwater rivers moving through them often carve a system of caverns and tunnels, such as the Paradise Ice Caves on Mount Rainier in Washington. These rivers deposit long ribbons of layered sand and gravel for miles along their tunnel's course. Eventually, when the stagnant ice sheet melts, these sand and gravel formations appear as landscape features called **eskers**, common in west-central Minnesota. Usually eskers are 30 or more feet (9 m) wide and just about as high, although they can be several miles long. Local depressions in the stagnating ice are sometimes filled with sorted sand and gravel washed out by the meltwater; then, when the ice mass disappears, cone-shaped hills called **kames** dot the landscape.

Long, streamlined hills called **drumlins** are also telltale signs of a glacier passage and they are common landforms. Aligned in the direction of ice flow, they can be used as natural compasses. Drumlins are made of glacial sediments, though they sometimes have a solid rock core. They have a steep slope on the end that

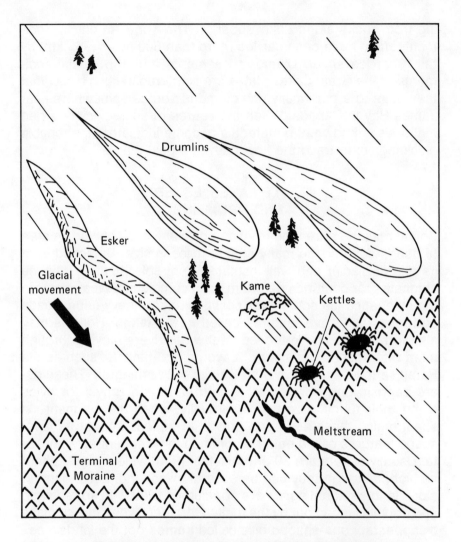

This diagram shows a composite of glacial features described in the text: drumlins, an esker, kettle lakes, a kame, and a terminal moraine being drained by meltwater.

faced the oncoming glacier and a long tapered slope on the opposite end. They are found in swarms of hundreds and can be seen in central Minnesota and near Weedsport, New York. Anyone flying into Boston can see several drumlins in the Boston Harbor area, the famous Bunker Hill being one.

Whalebacks are a near relative of drumlins. These are solid granite bedrock mounds with their shallow, sloped end pointing in the direction of the oncoming glacier and their blunt end pointing the other way, opposite from the orientation of drumlins. The blunt down-glacier end is formed by the ice "plucking" chunks of rock away as it grinds over the whaleback. These formations are common in eastern Maine, an especially large one being on Mount Desert Island.

If you are lucky enough to find a lake now being fed by glacial meltwater, you will be in for a special treat. Light rays penetrating the water are scattered by the fine silt particles of rock flour, making the water appear to be powder blue or turquoise. This coloring is caused by a process of light scattering similar to that which makes the sky appear blue. A spectacular example is Lake Louise in Banff National Park, Alberta, Canada. The lake presently is being fed by meltwater from the Lefroy and Victoria glaciers high on the Continental Divide.

If you ever travel up a fjord—perhaps along the Alaskan and Norwegian coasts, where fjords are so numerous—keep in mind that you are riding high over a glacial valley floor carved out by ancient ice to awesome depths. Norway's Songe Fjord near Bergen extends 120 miles (192 km) inland and plunges to a depth of nearly 4,000 feet (1,200 m) below sea level!

The great ice comes and goes. We now seem to be near the end of a global-warming period. After a brief time, geologically speaking, the long-range climate forecast calls for the great ice sheet once again to advance. When it does it will erase the past and reshape the land anew, as it has done at least seven times over the past 700,000 years.

5

ICE AGE ANIMALS

GAME HUNTERS OF 11,000 YEARS AGO

Standing on the exposed sandy bottom of a stretch of Aziscohos Lake in the western mountains of Maine, Richard Gramly, curator of anthropology at the Buffalo Museum of Science, described what it was probably like 11,000 years ago when the first settlers, Paleo-Indians (meaning "Indians of ancient times"), lived there. Gramly and his student diggers were on an archaeological expedition to uncover the Paleo-Indian site. "This part of the lake was the shore of an ancient river carved by glacial ice," said Gramly. "Grasses, sedges, dwarf huckleberries, low flowering herbs, and lichens grew among the rocks. There may have been a few small stands of spruce and other northern trees in sheltered locations. It was part tundra.

"This was good caribou country at one time," Gramly continued. "And that's why these people came here—to find meat for food, hides for clothing, and bone for tools. Maybe it was a spring or summer site visited seasonally by people who moved up from country farther south."

Gramly and his student workers discovered a killing ground, a place where hunters ambushed herds of caribou and other large animals moving up and down the valley. Ten spear points, four of which were unbroken and as sharp as the day they were made, were excavated at the kill site. This discovery and the associated campsite make the ancient settlement unique in eastern North America.

The people who came here 11,000 years ago were from Asia. Their skin was copper in color, their eyes were dark above wide cheekbones, and their hair was straight and black. Their ancestors were people who had trod some 55 miles (89 km) across a broad bridge of land linking Siberia to Alaska before the close of the last glacial period. This land bridge—the exposed floor of the Bering and Chukchi seas—was "open" from about 25,000 to 14,000 years ago, when the sea level was about 300 feet (90 m) lower than it is today because large amounts of water were locked up as ice.

Because this broad northern region receives relatively little moisture, it escaped the ice so plentiful elsewhere. At the close of the great ice age, Alaska and the northern Yukon were an Arctic paradise, harboring dozens of species of big game.

GIANTS OF THE ICE AGE

Fossils from many parts of North America are evidence that many large beasts roamed the land, most of which became extinct after the ice began to melt. There were horses, giant bison, giant beavers, giant ground sloths, bears, armadillos, woolly mammoths, mastodons, yaks, musk-oxen, woolly rhinoceroses, and saber-toothed cats.

Why so many giant animals? According to biologists, a cold climate favors large animals, since large animals lose body heat at a slower rate than do smaller animals. Although some of these animals, such as ground sloths and armadillos, came northward into the southern United States from South America, others, such as horses, saber-toothed cats, mammoths, antelopes, and musk-oxen, crossed over the land bridge from Asia.

What happened to all those wonderful giants? One of the mysteries of the last ice age is why so many of the large animals died out only a few thousand years after the retreat of the ice (although some had become extinct earlier). Most had died out 8,000 years ago, about 5,000 years after waves of Paleo-Indian settlers entered North America from Asia. Gone were the elephants, camels, horses, ground sloths, mastodons, giant beavers, dire wolves, and saber-toothed cats.

Why such a great period of dying out? Some biologists feel that skilled hunters among the Paleo-Indians killed off many of the giants, including the mammoth, bison, and horse. Archaeologists have come upon killing grounds such as those in Colorado, revealing the bones of hundreds of animals. Large groups of hunters apparently stampeded the animals and drove them into a confined area, where the animals were killed with stone-pointed spears and then butchered. Biologists also think that some species might have become extinct due to changes in the climate. For example, mastodons were native to spruce forests. But with the retreat of the ice came a dry period, and many of the spruce forests were replaced by pine and hardwoods. Such a change in the environment also reduced the numbers of the great beasts until there were too few left for a population to survive. All the "hugest, and fiercest, and strangest forms," as one biologist described them, disappeared almost overnight.

6

WHY ICE AGES COME AND GO

THE SUN OR THE EARTH ITSELF AS THE CAUSE?

No one knows for certain what triggers an ice age, but climatologists have some ideas. All their theories fall into one of two general groups: (1) ice ages are set off by astronomical events, for example, a decrease in the energy output of the Sun, or the solar system passing through a large cosmic cloud of space dust; and (2) ice ages are caused by events on the Earth itself, for example, changes in the ocean currents or times of widespread mountain-building.

The Isthmus of Panama, linking North and South America, uplifted about 3.5 million years ago. When this huge land bridge was thrust up it blocked the warm currents flowing westward from the Atlantic into the Pacific Ocean. Thus, an increased amount of warm water was forced into the North Atlantic Ocean toward Newfoundland and Greenland. This larger surface area of warm water would have led to an increase in evaporation and could have resulted in increased precipitation in the form of snow—enough snow to support the growth of ice fields.

Some climatologists think that the northern ice sheets are controlled by the Gulf Stream. According to this theory, as that great current of warm water shifts its position (see page 5) from time to time and becomes stronger or weaker, it causes an alternating spread and retreat of northern ice.

The ocean-currents theory sounds plausible, but it is still only a theory. It assumes that once a massive buildup of ice begins, there is a gradual decrease in precipitation, with the eventual result that the glacial region is deprived of snow and stops advancing. It stops because its rate of melting is now greater than its rate of growth. This, then, is the dry stage in climate change, a time when large amounts of loess enter the air. The fine mineral dust settles in thin layers on the continental ice sheets as it slowly falls out of the air, causing a sharp drop in the amount of the Sun's energy reflected back into the atmosphere by the ice. The result is that the increased absorption of heat caused by the loess particles speeds melting. Although this may occur in certain places, it certainly did not occur in Antarctica and occurred only to a very small extent in Greenland. You can demonstrate this principle of melting by lightly sprinkling wood ashes over a patch of snow and then observing the difference in the rate of melting between your darkened patch and the surrounding, undisturbed snow.

CHANGES IN ENERGY FROM THE SUN

Several scientists are convinced that ice ages come and go during periods when the Earth receives lesser or greater amounts of energy from the Sun. As the Earth orbits the Sun, its distance from the Sun changes. This happens for a number of reasons. One is that the Earth's orbital path around the Sun is not a perfect circle, but a slightly elongated one called an ellipse. It now seems quite likely that changes in the Earth's orbit have been the major cause of the advancing and retreating of ice over the past 2 million years.

The Earth reaches its greatest distance from the Sun once every 93,000 years. When at its greatest distance, the Earth receives about 20 percent less energy than when we are closest to the Sun. This is a large change in energy when we consider that a drop of only 13 percent would bring on a super ice age—one covering the Earth's entire surface with a blanket of ice 1 mile (1.6 km) thick. On the other hand, a rise of 30 percent would bring on a heat wave that would destroy virtually all life on the Earth.

When the Earth is closest to the Sun, we might expect the reverse to occur—a widespread melting of the ice. Scientists generally agree that the rapid end of the last great ice age is best accounted for by a combination of two factors: (1) an orbital change that brought the Earth closer to the Sun; and (2) a slight change in the amount by which the Earth is tilted on its axis. Somewhat less of a tilt would make it "unusually" warm at mid-latitudes, near where the lower edge of the great ice sheets were.

Astronomer John A. Eddy is convinced that glaciers come and go in pace with decreases and increases in the activity of the Sun. This has nothing to do with the Earth's changing distance from the Sun. Rather, it relates to the fact that the Sun does not provide us with a steady output of energy. In fact, there seem to be cyclical periods of high and low output that occur with some regularity. Eddy says that every rise in solar activity matches a time of glacial retreat. He concludes that his "early results in comparing solar history with climate make it appear that changes on the Sun are the major agent of climate changes lasting between 50 and several hundred years."

CLIMATE FORECAST:
COLDER, WITH ICE

Physicist Willi Dansgaard sees two such solar energy cycles, one peaking about every 78 years, the other peaking about every 181 years. His long-range forecast is for a gradual warming beginning in the late 1980s and lasting until the year 2015, which will bring us back to the average world temperature of 1960. Then it will start getting cold again, and still colder, for the next 50 years or so. The long-long-range forecast over the next 20,000 years is toward extensive Northern Hemisphere glaciation and a cooler climate.

7

DATING EVENTS IN THE EARTH'S PAST

In several parts of this book we have said that scientists tell us that this or that ancient event—an ice age or a time of drought—took place so many thousands or millions of years ago. How do they know? Estimates of the age of the Earth's various rock and fossil samples are based on various methods of dating, a few of which will be briefly described here.

TELLING GEOLOGIC TIME WITH ATOMIC CLOCKS

Around 1900, scientists learned that atoms of certain elements said to be radioactive "decay" (that is, lose some of their sub-atomic particles) at certain rates and turn into a different kind of atom, resulting in a new element. For instance, uranium slowly changes into lead, and potassium slowly changes into argon. This happens because radioactive atoms are big and have many subatomic particles. The number of subatomic particles makes the atom unstable, meaning that the atom loses some of its particles. It is this loss of subatomic particles that makes some of the atoms change (decay) into atoms of a different element. The scientist dating a rock sample compares the number of unchanged uranium atoms with the number of atoms in the sample that have changed. The amount of time needed for half of the atoms of a radioactive element to change is called its **half-life**.

Nothing seems to affect the half-life of any radioactive element—not even changes in temperature or pressure. Since scientists know the half-life of a given radioactive element, and since they can compare the number of new (lead) and old (uranium) atoms, they can then tell how long the radioactive "clock" has been running.

Different radioactive elements have different half-lives. Here are three different radioactive elements that are used to tell the age of rocks, the elements they change into, and their half-life.

this radio-active element	changes to	and has a half-life of
uranium 238	lead 206	4,510 million years
potassium 40	argon 40	1,300 million years
rubidium 40	strontium 87	47 million years

It is because each radioactive element has its own private rate of decay that it can be used as an atomic clock. Thus atomic clocks give scientists an accurate way to say that one rock, for example, is 170 million years old as opposed to another that is 30 million years old. But what about dating objects or events more recent in time?

DATING ONCE-LIVING REMAINS

Since all living matter known to us contains carbon, this element can be used as an atomic clock to date the remains of once-living materials. Ordinary carbon is in the form of carbon-12 (the number 12 tells us how many of certain particles are contained in the nucleus of each atom). The form of carbon used as a short-term atomic clock is carbon-14, which has a slightly more massive nucleus than carbon-12. Carbon-14 is continuously being produced in the atmosphere out of nitrogen-14. This happens as the nitrogen-14 is bombarded by energetic particles from space called **cosmic rays**. The leaves of green plants continuously take in carbon-14 right along with carbon-12 as they take in carbon dioxide. So green plants, and all the organisms that depend on them for food—which includes almost all animals—take in carbon-14. But carbon-14 does not remain carbon-14 once it is part of the organism. Instead, it changes back into nitrogen-14.

When a plant or animal dies, it stops taking in carbon-14, so the amount of carbon-14 an organism contains immediately begins to decrease the moment the organism dies. When an old bone, shell, or other piece of once-living matter being dated is analyzed, the scientist compares the number of carbon-14 atoms to the number of carbon-12 atoms. That comparison, or ratio, reveals the age of the once-living matter. Since carbon-14 has a half-life of only 5,730 years, it is useful only for short-term dating—as compared with, for example, the uranium-lead atomic clock.

The carbon-14 method has proved extremely useful in dating wood, charcoal, peat, bones, marine shells, and other once-living matter. Scientists make two assumptions when they use carbon-14 as an atomic clock: (1) that all organisms take in carbon-14 at a more rapid rate than the rate at which the carbon decays into nitrogen-14 in their living tissues; and (2) that the rate of carbon-14 production in the atmosphere is fairly constant, as is the ratio of carbon-14 to carbon-12.

THE TALES TOLD BY
TREE GROWTH RINGS

Once scientists have dated what is left of an ancient fossil tree, for instance, by the carbon-14 method, they can examine the

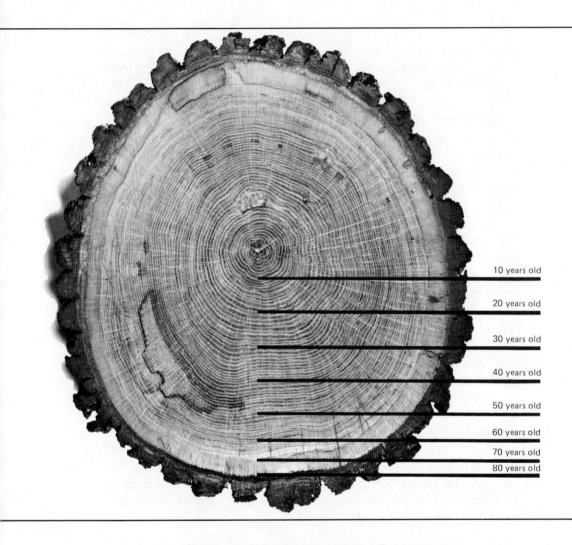

10 years old

20 years old

30 years old

40 years old

50 years old

60 years old

70 years old

80 years old

By comparing the thickness of a tree's annual rings, scientists can find out about the rainfall pattern during the time the tree was alive. They also can tell the tree's age by counting the number of growth rings.

tree's annual growth rings to discover what the rainfall pattern was during the time the tree was alive. Or, if the tree is still alive, they can determine its age by counting the growth rings. California's giant Sequoia trees reveal ages up to 4,000 years, and bristlecone pines to about 5,000 years. Some fossil trees containing tree rings go back even further in time, to about 8,000 years ago.

In some geographical areas, there is a close relationship between the temperature, the amount of precipitation in a given year, and the width of a tree's annual growth ring for that year. Considerable precipitation and warm temperatures mean rapid growth and relatively wide growth rings; a cool season with little rainfall means little growth and relatively narrow growth rings. Tree-ring growth in the interior plains and plateaus of the United States and Canada reveals some interesting climate changes that have occurred over the past 850 years. For example, during nearly all of the thirteenth century the climate was dry, but then the fourteenth century had a stormy climate. The last quarter of the sixteenth century—the climatologists tell us—was the driest period in the last 650 years.

VARVES AS CLUES TO PAST CLIMATES

Another short-term method of deciphering past climates is by studying samples called **varves,** the built-up layers of silt and clay laid down on the bottoms of lakes and ponds year by year. Varves provide climate records going back only a few thousand years. And the bodies of water being studied must have been subject to seasonal freezing and thawing.

In summer all kinds of materials, including dust carried by the atmosphere, enter a lake and slowly settle to the bottom. In winter, however, the surface ice prevents foreign matter from entering the lake. Only the fine clays already suspended in the water settle out and sink to the bottom. A long hollow tube pushed deeply into the soft bottom of a lake can bring up twenty or more varve layers, each one representing an equal number of years' deposits. No two years produce quite the same thickness of varve layering. The thickness of a varve indicates the amount of silt deposited during a given summer, which in turn depends largely on the strength of the spring thaw and resulting flooding. Heavy rainfall during a summer will cause relatively large amounts of soil to be washed into a lake by runoff and so contribute to the thickness of varves.

Varve records also reveal quite accurately the dates of the recession of glaciers. This is because varves cannot form beneath glaciers any more than they can form beneath a lake bed permanently topped with a layer of ice. So a lake bed only recently exposed by the retreat of a glacier will have relatively young varves, as compared with an old lake bed long free of glacial activity. In Scandinavia, for instance, varve records go back about 13,700 years. Because the youngest varves are found in the north and the oldest ones are located in the south, it is an easy matter to trace the history of the last ice sheet as it retreated northward.

There are also fossil varves. These are formed as the soft materials forming a lake bottom or ocean bottom are gradually compressed and over millions of years turned to rock. At some later period, that rock is then thrust up as part of a new mountain. A geologist studying a sample of this fossil varve can read it in much the same way that one can read a fresh sample of varves removed from a lake bottom only yesterday. Scientists working in Alberta, Canada, have uncovered a fossil varve record 300 million years old and spanning a period of 900 years.

SOILS AS CLUES
TO PAST CLIMATES

Ancient soils also serve as climate indicators, but only of the past million or so years. As varves show a sequence of differing layers, so does a core sample of soil. One such soil core taken in Czechoslovakia reveals the sequence of alternating warm and cool periods for nearly a million years. One of the clues is the presence in certain of the soil layers of tiny snails known to prefer colder climates.

EVIDENCE FROM POLLEN

Rocks formed out of the clay, sand, and other materials that rain down onto a lake bottom or ocean floor are called **sedimentary rock**. These are the rocks that contain fossils. Among the many different materials forming the sediments are countless billions of grains of plant pollen produced each growing season and carried far and wide on the wind. Since each kind of plant produces its own particular kind of pollen, scientists can tell what groupings of plants grew at this or that time in the Earth's history. Fossilized pollen and the various changes in types of plant groupings found in different locations are evidence that there have been a number of distinct changes in climate over the past 30,000 to 35,000 years.

THE RISE AND FALL
IN SEA LEVEL

One way we can trace the shorelines of ancient seas is to trace the extent of fossilized ripple marks and other such characteristics of old sea bottoms. Such studies reveal not only the locations of ancient seas but also the rise and fall of shorelines. This is a very reliable indicator of climate change because a change in sea level tells us that something is happening to the ocean water. A lowering of the mean sea level would indicate that water is being removed to make glaciers. A rise would indicate that glaciers were going through a stage of melting. As the graph indicates, about 17,000 years ago the mean ocean level was about 300 or more feet (90 m) lower than it is today. So at that time, there must have been extensive glaciers covering the continents, some up to 2 miles (3.2 km) deep.

We have hardly touched on the many methods scientists use to decipher the coming and going of ice ages and other climatic events that have taken place during the Earth's long history. One of the many lessons scientists have learned from studying the Earth's past is that whatever has occurred in the past can be used as a key to predict the future.

8

VISITING MOUNTAIN GLACIERS

To get ready for the next age of ice you might find it helpful to visit one or more present-day mountain glaciers during your next vacation. If you decide to do this, you should know what you may be getting into. Although the national parks listed in this chapter have well-marked and safe trails leading to many glacial sites, and the park rangers permit climbing expeditions, you would be well advised to have an experienced guide accompany you. Be sure to bring suitable climbing equipment.

There are two major sources of information about the locations of active glaciers and scarred landforms showing that glaciers passed this way long ago. You can write or call your state's geological survey, or conservation department, and ask where telltale signs of ancient glaciers may be found in your state. Or, for detailed information about the glaciers and glacial remains in the national parks listed, write the park superintendents. Some states have state parks containing evidence of glaciers—Glacial Lake State Park in Minnesota is one example. When you study old glacial remains, see if you can tell in which direction the glacier advanced. Can you identify erratics? Look for drumlins, whalebacks, and other prominent surface features. If you come across a terminal moraine, see how many of its materials you can identify.

GLACIER NATIONAL PARK. In Montana, in the northwestern part of the state. The park sprawls over a million acres, has 70 miles (113 km) of park roads and 1,000 miles (1,609 km) of hiking trails. It is the fourth largest of the national parks. One of the park's most spectacular views is from the top of the Garden Wall, a rugged climb of some 1,500 feet (460 m) in just under one mile (1.6 km). For information, write the Park Superintendent, Glacier National Park, West Glacier, Montana 59936.

MOUNT MCKINLEY NATIONAL PARK. In Alaska, about 120 miles (193 km) southwest of Fairbanks, the park is second only to Yellowstone in size and spreads over almost 2 million acres. The major attraction is the heavily glaciated mountain itself, the highest peak in North America, rising to 20,320 feet (6,194 m). The largest glaciers are on the mountain's south side and are valley glaciers. In 1956, the mountain's Muldrow Valley Glacier began a rapid advance after a long period of slow recession. One of the largest valley glaciers on the mountain, it advanced almost 4 miles (6.4 km) in less than a year. It is some 35 miles (56 km) long. The mountain's Kaliltna Glacier is 45 or more miles (72 km) long. Most of the park's streams and rivers are fed by melting glacial ice. The park usually is open June 1 to September 10. For information, write the Park Superintendent, Box 2252, Anchorage, Alaska 99051.

MOUNT RAINIER NATIONAL PARK. In Washington, about 55 miles (89 km) southeast of Tacoma, the park covers almost 25,000 acres. The chief attraction is Mount Rainier itself, one of North America's loftiest and most scenic mountains, which rises to 14,410 feet (4,392 m). This mountain glacier wonderland has eleven main glaciers, averaging 4 to 6 miles (6.4 to 9.7 km) long.

The view here from Mount McKinley National
Park is to the southwest toward peaks around
the source of the Tokositna Glacier.

Mountain glaciers sometimes have great hollows called ice caves. Here a party of climbers on Mount Rainier is about to enter an ice cave.

The total of twenty-eight glaciers makes Rainier one of the most heavily glaciated single peaks in the United States. The mountain is what remains of a huge volcano. Paradise Glacier is the most accessible and offers some beautiful ice caves that open in late summer. Emmons Glacier is the mountain's largest and flows down the northeast flank. Be sure to visit Nisqually Glacier and examine the large amount of glacial drift dumped as a terminal moraine. Mount Rainier also has many glacial lakes. The park is usually open from May to October. For information, write the Park Superintendent, Longmire, Washington 98397.

OLYMPIC NATIONAL PARK. In Washington, tucked away in the northwest corner of the state, the park ranges over almost 900,000 acres. Glacier-coated Mount Olympus is in the heart of the park and rises to 7,965 feet (2,428 m). If you explore the glaciers on this mountain keep an eye out for large granite erratic boulders carried down from the mountains of western British Columbia during the last great ice age and deposited 3,000 feet (900 m) up the slopes of Mount Olympus. At least sixty glaciers extend their icy fingers down the Olympic Mountains today. Six on Mount Olympus are still active, since the mountain receives up to 140 feet (43 m) of snow a year. The six are Hoh, Blue, White, Hubert, Jeffers, and Hume glaciers and are up to 2 miles (3.2 km) long. Most of the larger glaciers of the Olympic range are on the northern slopes, where they are protected from the Sun. The park is open year-round. For information, write the Park Superintendent, 600 East Park Avenue, Port Angeles, Washington 98362.

YOSEMITE NATIONAL PARK. In California, about 150 miles (240 km) east of San Francisco on the western flank of the Sierra Nevada, the park spans an area of over 750,000 acres. There are many reminders of the last great ice age here. Among them are many small hanging valleys now high above the main valley floor and containing evidence of tributary glaciers that once fed the

An aerial view of the Hoh Glacier on Mount Olympus. Facing page: Hikers climb their way up a cliff face on one of Yosemite National Park's many trails for tourists visiting the park.

main valley glaciers. There are also clear mountain lakes fed by glacial meltwater and relatively fresh moraines with an outwash of gravel, sand and silt extending downstream. The most spectacular attraction is Yosemite Falls, one of the world's highest free-leaping falls. It flows over the lip of a hanging valley and plunges a total of 2,425 feet (739 m). It is more than nine times the height of Niagara Falls. The falls are best viewed from Glacier Point, 3,200 feet (970 m) above the valley floor. Other free-leaping falls include Bridalveil and Illilouette falls. The park is open year-round. For information, write the Park Superintendent, Box 577, Yosemite National Park, California 95389.

GLOSSARY

Climate. A region's weather averaged over a long span of time. From the Greek word *klima*, meaning "slope or incline," and referring to the degree of slant of the Sun's rays relative to the Earth's surface.

Climatologist. A scientist who studies climate and climate changes.

Cosmic rays. High-speed subatomic particles from space and the Sun. They can penetrate 3 feet (0.9 m) of lead or 3,000 feet (900 m) of water. The Earth's atmosphere protects us from these damaging particles by breaking them up.

Crevasse. A deep trench-like opening in glacial ice.

Drumlin. A long and narrow mouse-shaped mound of sand, gravel, and clay deposited by glacial ice. Drumlins point along the line of glacial flow, their blunt end pointed up-glacier and their gently-sloping tails pointed down-glacier. A drumlin may be as much as half a mile (0.8 km) long and 100 feet (30 m) high.

Erratic. A boulder or other rock that has been carried by a glacier sometimes over hundreds of miles from its source. An erratic may be identified as such because often it is a different type of rock from the bedrock on which it rests.

Esker. A winding ridge of sand, gravel, clay, and other materials carried and deposited by streams that run under or through glacial ice. Eskers form in layers (which is not so of a drumlin). They may be up to about 50 feet (15 m) high from a fraction of a mile long to 100 miles (161 km) long.

Fjord. A steep-walled, water-filled valley gouged out by glacial ice flowing down to the sea. There are many fjords along the coast of Alaska and Norway.

Firn. Granular ice that has recrystalized from snow. In form it is between snow and glacial ice.

Geologic time. The portion of time occurring before writing was invented and written records were kept.

Glacial drift. Loose rock and rubble created by an advancing glacier.

Glacier. A large mass of ice formed by a long accumulation of snow. Under the influence of gravity, glaciers flow down slopes and are kept in motion as increasing amounts of snow are added to the source region of the glacier.

Glaciologist. A scientist who studies glaciers.

Gondwana. A continent formed during the Earth's early history, when a large supercontinent broke into a southern half and a northern half, the former of which was called Gondwana.

Gulf Stream. A current of relatively warm water flowing up the eastern coast of the United States and across to Britain and farther north. The Gulf Stream branches near Britain, traveling southward and then westward, forming a closed loop.

Half-life. The length of time needed for half of the atoms of a radioactive element to change, resulting in a different element. For example, it takes 4,510 million years for half of the uranium atoms in a rock sample to change into lead.

Ice age. Any extended period of time during which a substantial portion of the Earth's surface is covered by "permanent" ice. There have been seven known major ice ages during the past

700,000 years, with the last one reaching its peak about 18,000 years ago.

Iceberg. A massive piece of ice, only a small part of which is visible above the surface of the water. Icebergs are commonly formed as large pieces of ice break off the ends of glaciers that flow into the sea, in Alaska and Greenland, for example.

Icecap. An ice sheet such as that covering Greenland and Antarctica. An icecap builds as a result of a steady accumulation of snow.

Interglacial period. The period of time between two succeeding ice ages. An interglacial period may last for 10,000 years or more.

Kame. A hill with steep sides, composed of sand, gravel, clay, and other glacial debris that has been deposited in layers. Kames differ from drumlins by lacking a characteristic shape and by being composed of layered debris.

Kettle lake. A lake hollowed out by glacial ice and formed by the melting ice.

Lateral moraine. A ridge of rock and soil debris that forms along the edges of a valley glacier.

Laurasia. A continent formed during the Earth's very early history, when a large supercontinent broke into a southern half and a northern half, the latter of which was called Laurasia.

Little Ice Age. That period from roughly 1400 to the mid-1800s, during which mountain glaciers advanced and a number of severe winters occurred.

Loess. Silt with smaller amounts of very fine sand and/or clay that may be carried by the wind and deposited on glacial ice and other surfaces.

Medial moraine. A ridge of rock and soil debris formed when two valley glaciers merge and their lateral moraines join.

Mountain glacier. A glacier that forms in the mountains and that may grow into a sprawling ice sheet which, when conditions are right, flows over the surrounding land.

Outlet glacier. A glacier that has advanced to the edge of the coast and is dumping, or "calving," enormous chunks of ice into the sea.

Pangaea. A large supercontinent that, during the Earth's very early history, broke into a northern part, Laurasia, and a southern part, Gondwana.

Rock flour. Rock crushed and pulverized by the action of glacial ice and carried by streams into glacial lakes.

Sedimentary rock. Fossil-bearing rock that was formed out of clay, sand, and other materials that fell to the ocean's or lake's bottom.

Striations. Small, parallel scratches in rock, indicating that the rock was once covered by a glacier.

Terminal moraine. Sand, gravel, rock, clay, and other debris deposited at the leading edge of a glacier and marking the farthest reaches of the glacier's advance.

Valley glacier. A glacier that flows along the floor of a valley.

Varves. Built-up layers of silt and clay laid down on the bottom of lakes and ponds year after year.

Whaleback. A near relative of drumlins, these are solid granite bedrock mounds with their shallow-slope end pointing up-glacier and their blunt end pointing down-glacier, opposite from the orientation of drumlins. The blunt down-glacier end is formed by the ice "plucking" away chunks of rock as it grinds over the whaleback. These formations are common in eastern Maine.

INDEX

Mount McKinley National Park, *13*, *25*
Mount Olympus, *54*
Mount Ranier National Park, 28, 31, 50, *52*
Mount St. Elias, *2*

New England, 22, 28, 33–34; *see also* United States
New Zealand, 20
North America, 20, 22–24, 27, 36
    climate change of, 6–7, 10, 21, 45
Northern Hemisphere. *See* Asia; Europe; North America

Ocean currents theory, 38
Olympic National Park, 53
Orbit of the earth, 38–39
Outlet glacier, 17

Pack ice, 10
Paleo-Indians, 34–36
Panama, Isthmus of, 37
Pangaea, 8–9
People, 34–36
Plants, 22, 47
Polar regions. *See* Antarctica
Pollen, 47
Precipitation, 14, 21, 37–38, 45

Radioactive clocks. *See* Atomic clocks
Radioactive elements, 41–42
Rain. *See* Precipitation
Ripple marks, 47
Rock flour, 27, 28
Rocks, dating, 42

Sargasso Sea, 6
Sea level, 14, 20–21, 47–48
Sedimentary rock, 47
Sierra Nevada Mountains, 10

Snow. *See* Precipitation
Snow lines, 20
Soil, 46–47
Solar energy, 10, 21, 38–40
    cycles of, 39–40
South America, 20, 35
Southern Hemisphere. *See* Africa; Antarctica; Australia; New Zealand; South America
South Pole. *See* Antarctica
Steppes, 20
Striations, 27, *29*

Temperature
    average world, 7, 21, 40
    changes in, 3, 8, 21
    icebergs and, 21
    tree growth and, 45
    See also Climate changes
Tertiary Period, 6–10
Tributary glaciers, 53

United States
    climate changes, 6–7, 45
    parks, 2, 13, 25–28, 31, 49–56
U.S. Coast Guard iceberg census, 17, 20

Valley glacier
    growth of, 24, *25*, 27, 56
    movement of, 17, 50
Varves, 45–46
Vegetation, 20

Weather conditions. *See* Climate changes
Weddell Sea, 10
Whalebacks, 33
Winters, 3–4

Yosemite Valley National Park, *26*, 27, 53, *55*, 56

THE AUTHOR

Roy A. Gallant is an adjunct full professor of English at the University of Southern Maine and director of the University's Southworth Planetarium, where he creates planetarium shows and lectures to student and adult groups. He is a former editor-in-chief of the Natural History Press, of The American Museum of Natural History in New York City; executive editor of Aldus Books, Ltd., of London (a subsidiary of Doubleday); and managing editor of *Scholastic Teacher* magazine. Currently, among his many other activities, he is serving as an earth science consultant for the magazine *Science and Children*, published by the National Science Teachers Association, and is a member of the New York Academy of Sciences and a Fellow of the Royal Astronomical Society, London.

Professor Gallant is the author of more than fifty science text and trade books for young readers and adults. For Franklin Watts, he has authored the critically acclaimed *Once Around the Galaxy* and is currently working on several titles in the area of computer applications.